IS BIOLOGY WOMAN'S DESTINY?

by Evelyn Reed

PATHFINDER
NEW YORK   LONDON   MONTREAL   SYDNEY

Copyright © 1972 by Pathfinder Press
All rights reserved

ISBN 978-0-87348-258-5
Library of Congress Control Number 2009930906

Manufactured in the United States of America

First edition, 1972
Second edition, 1985
Fourteenth printing, 2018

PATHFINDER
www.pathfinderpress.com
E-mail: pathfinder@pathfinderpress.com

CONTENTS

Introduction   5

Is biology woman's destiny?   9

Notes   39

Introduction

WOMEN'S RIGHTS ARE under severe attack today by the government, employers, and right-wing forces.

Abortion, a right won through struggle by women at the beginning of the 1970s in the United States, is being restricted more and more by federal, state, and local legislation and court rulings. Right-wing outfits are carrying out a wave of abortion clinic bombings in an attempt to intimidate women from obtaining this simple medical procedure under safe conditions.

Democratic and Republican politicians let the federal Equal Rights Amendment go down to defeat in 1982.

Social spending cutbacks have led to the closure of many of the already insufficient facilities for child care in this country.

Affirmative action gains to ensure women the right to the same opportunities as men in hiring and on the job are being eroded. The government and bosses continue to relegate women to lower-paid positions.

Employers promote sexual harassment on the job with the aim of keeping women workers intimidated and docile. Maternity leave benefits are being taken away.

This is all part of the bosses' overall assault against the

rights and living standards of working people and the oppressed.

The aim of the government-employer offensive against women is not to push them out of the work force, but to force women to accept lower wages and substandard conditions. The capitalists seek to convince both male and female workers that women are not "real" workers anyway; they are just earning "pin money," a "second income," and so on. If a woman gets laid off, the employers claim, this should not count in the overall unemployment statistics the same way as the layoff of a man—the "main breadwinner."

The target is women as a sex—those who are currently working, and those who are not. At a time when the economic and social pressures on working people and their families are intensifying, the capitalists are also cranking up their propaganda campaign to justify women's second-class status.

With more than 50 percent of women working today, a large number of them single mothers, the rulers' propaganda cannot simply rely on the reactionary old slogan that "a woman's place is in the home." It's OK for a woman to work, they say, but her main responsibilities remain housework, husband, and family. The glories of motherhood are extolled. Women work two shifts; the second one begins when they come home from the job.

The big-business press, the churches, and the schools all help to disseminate this antiwoman propaganda. The rulers also call on the services of scientists, who misuse their credentials to cloak reactionary social and political prejudices in pseudoscientific garb. Recently, for example, a prominent scientist, speaking about differences between men and women, wrote, "it is perfectly good biology that business and profession taste sweeter to [men], while home and child care taste sweeter to women."

Bogus biological and anthropological theories such as this are peddled to "prove" that genes determine the social roles of men and women today—and that they always have and always will. Such pseudoscientific propaganda is used not only to rationalize the oppression of women, but also to propagate racism, to claim that class inequality is part of the "natural order" of things, and to justify biblical creationism against a scientifically grounded understanding of the origins of humanity and its evolution.

Pathfinder Press is reissuing this 1972 pamphlet by Evelyn Reed as a tool to help all those who are fighting the sharp blows against the rights of women. It provides a powerful answer to one of the fundamental arguments underlying all of the sexist propaganda—that biology is woman's destiny.

Evelyn Reed, who died in 1979, was a leader of the Socialist Workers Party for many years, and a prominent spokesperson for the feminist movement in this country in the late 1960s and early 1970s. She was a founder of the Women's National Abortion Action Coalition (WONAAC), an organization that played an important role in the fight to legalize abortion leading up to the 1973 Supreme Court decision. Reed is also the author of many articles, pamphlets, and books on women's liberation.

Her major work, *Woman's Evolution,* develops more extensively the themes introduced in this pamphlet. Since its publication in 1975, *Woman's Evolution* has been a source of education and inspiration to many feminists and supporters of women's rights in this country and worldwide. It has been translated and published in French, Spanish,

Swedish, Turkish, and Farsi.

Two other books by Reed—*Problems of Women's Liberation* and *Sexism and Science*—are also published by Pathfinder Press, and Reed is the author of the introduction to Pathfinder's edition of *Origin of the Family, Private Property, and the State* by Frederick Engels.

Sonja Franeta
June 1985

Is biology woman's destiny?

MANY WOMEN IN the liberation movement, especially those who have studied Engels's *Origin of the Family, Private Property and the State,*[1] have come to understand that the roots of women's degradation and oppression are lodged in class society. Quite correctly they have coined the term "sexist" to describe the capitalist social system, the final stage of class society, which discriminates against women in every sphere of life.

What women remain unsure about, however, is whether or not their biology has played a part in making and keeping them the "second sex." Such uncertainty is quite understandable in a male-dominated society where not only is history written by those who uphold the status quo but all the sciences are likewise in their hands. Two of these sciences, biology and anthropology, are of prime importance in understanding women and their history. Both are so heavily biased in favor of the male sex that they conceal rather than reveal the true facts about women.

Perhaps the most pernicious pseudoscientific propaganda on female inferiority is that offered in the name of biology.

According to the mythmakers in this field, females are biologically handicapped by the organs and functions of motherhood. This handicap is said to go all the way back to the animal world and makes females helpless and dependent upon the superior male sex to provide for them and their young. Nature is held responsible for having condemned females to everlasting inferiority.

It is obvious that females are biologically different from males in that only the female sex possesses the organs and functions of maternity. But it is not true that nature is responsible for the oppression of women; such degradation is exclusively the result of man-made institutions and laws in class-divided patriarchal society. It did not exist in primitive classless society and it does not exist in the animal world.

This falsification of natural and social history has been propagated to exonerate a sexist society. The oppression of women is justified on the ground of their biological makeup. The implication is obvious: Why should women fight against their oppression and seek their liberation when their troubles flow from their genetic makeup? What good will it do to change society if women cannot change their biology? This theme is drummed into our heads by every means available, from the cradle on. To believe the male supremacists who pose as scientists, biology is woman's destiny and she had better recognize and submit to it.

In truth, it is no less false to say that biology is woman's destiny than to say that biology is man's destiny. This reduces humans to the animal level. For if women are nothing but breeders then men must be nothing but studs. Such a reduction leaves out the decisive distinctions between humans and animals. Humans are above all *social* beings who have long since separated themselves from their animal origin and conditions of life. To understand the differences between the sexes, let us first examine these distinctions between humans and animals which make humanity a totally new and unique species.

Humans a unique species

Ever since Darwin demonstrated that humanity arose out of a branch of the higher apes, numerous studies have been made showing the similarities between humans and animals. But there are all too few studies showing what is even more important—the enormous distinctions between humans and animals that make us a unique species standing above and beyond all forms of animal life.

The central source of this uniqueness has been pinpointed by the Marxists. It is the capacity of humans to engage in labor activities and produce the necessities of life. No animal species does that. This "labor theory" of human origins was first set forth by Engels in his essay "The Part Played by Labour in the Transition from Ape to Man."

Today such leading authorities in archaeology and anthropology as Sherwood Washburn, William Howells, Kenneth Oakley, V. Gordon Childe, and others use toolmaking as the criterion that distinguishes humans from animals. As Washburn sums this up, "It was the success of the simplest tools that started the whole trend of human evolution and led to the civilizations of today."[2] Gordon Childe sustained Engels's thesis when he said, "Prehistoric archaeology shows how man became human by labour . . ."[3]

Those who downgrade labor activities often contend that apes likewise use natural objects as tools and therefore labor cannot be taken as the basic factor that brought about the humanization of our species. But the point is, no matter how clever a primate is in the use of its hands—and in captivity they can be taught to do quite a number of things—no animal species, including primates, is capable of becoming a toolmaker. There is no division of labor between the sexes among primates and no prehuman species depends upon systematic labor activities for its survival. "Hands with which to pluck and arms with which to convey the edible plunder to the mouth suffice. This is the technique of our anthropoidal relatives," says E. Adamson Hoebel.[4]

By contrast, humans are so completely dependent upon labor activities that should this productive capability cease we would soon perish as a species. Labor activities, therefore, have brought about a new mode of survival and development for a unique species: humankind. We are not merely reproducers but producers of the necessities of life.

The importance of production can be seen in the drastic change that labor activities brought about in the relations between humans and nature. Fundamentally, animals are the slaves of nature, subject to biological forces and processes over which they have no control. Humans, on the other hand, have reversed this relationship. Through labor activities humans have brought nature under their influence. In other words, one species, humankind, not only became liberated from direct biological control but even became the controller of its former dominator, nature. As this is sometimes put, the history of animals is made for them, but humans alone make their own history.

Along with this mastery over nature, humans also began to cultivate new needs, which is another characteristic absent from the animal world. While animals are limited to satisfying the same old natural needs for food and procreation, humans have developed an endless series of new needs, all of them higher (at least in the sense of being more sophisticated) cultural needs. To take a few examples from the technical realm: out of the first fist-axe there arose the need for an axe with a handle on it. From the crude digging-stick there came the need for and invention of the plough. The simple loom and spinning wheel led on to the complex textile industry. Construction needs passed beyond thatched huts to the building of factories and skyscrapers. The oxcart was left behind as new needs for rapid transit after the industrial revolution led to the train, the automobile, the jet plane, and the spaceship.

Cultural needs of all types, in education and the arts and sciences, arose as part of the new activities and relations of humans in social life. Even the basic biological needs for food and sex became altered and reshaped in human life. Humans do not eat, mate, or procreate as animals do, but in accordance with their own changing cultural standards. As Marx wrote, "Hunger is hunger; but the hunger that is satisfied with cooked meat eaten with fork and knife is a different kind of hunger from the one that devours raw meat with the aid of hands, nails, and teeth."[5]

Along with the vast change made by humans in external nature, their environment, they made equally important changes in their own internal nature. It is often pointed out that in their physical appearance humans shed their hairy coat and other ape-like characteristics. But even more important was the shedding of their former animal reactions, which were replaced by a humanized social nature. Today we have lost virtually all the animal instincts we started with—these have been displaced by learned behavior.

This brief review of some of the vital distinctions between humankind and all other species refutes the thesis that humans are "nothing but" animals with a few extra tricks. It is far more accurate to say that, while we still share certain biological characteristics in common with the animals, we have raised ourselves far above their limited existence. We have been formed and transformed in and through our own productive activities and social forces, so that we are no longer the slaves of our biological makeup. As the Michigan anthropologist Marshall Sahlins puts it, "The liberation of human society from direct biological control was its great evolutionary strength.... Human social life is culturally, not biologically, determined."[6] This is the starting point for demolishing the myth that biology is woman's destiny. Beginning with the primary proposition in this propaganda, let us examine what I call—

The 'uterus theory' of female inferiority
Biology, like anthropology, is a young science and equally subject to misinterpretations, superficial conclusions, and downright lies in questions that have grave social and political implications. This makes it doubly difficult to uncover the truth about the female sex, since so many biologists as well as anthropologists are captives of capitalist ideology. They assume that because woman is born with a uterus she can never liberate herself from direct biological control and must forever remain enslaved to her procreative functions.

This "uterus theory" of female inferiority is no more valid than its corollary, the "penis theory" of male superiority. For some curious, unexplained reasons, these sexual-procreative organs are supposed to have determined all the other capacities of the sexes. Woman, rendered stupid by the functions of her uterus, was unable to develop her brains, talents, and higher cultural capabilities. Man, on the other hand, with his upstanding sexual muscle instead of the ignominious uterus, could develop his intellect and associated abilities. Both propositions are fiction, not science.

In actuality it is the male that is handicapped in the animal world, not the female. This is due to the disruptive characteristics of male sexuality in nature. As the record shows, males are highly competitive and fight other males for access to females. Although this is often called "jealousy," it is not jealousy in our sense of the term, namely, the desire to possess a particular female. Rather, it is a crude, pugnacious instinct, unmodified by any feelings of individual preference or tenderness, which drives the male animal to seek access to any and all females. In some species the males may fight one another merely for a place in the breeding grounds; in others they may fight even in the absence of females. As Sir Solly Zuckerman says, "The pugnacity of rutting animals is an expression of their physiological condition and not necessarily determined by the presence of females."[7]

Due to this combative characteristic of male sexuality, male animals are separatistic, individualistic, and unable to band together in mutually cooperative groups. At best, they are able to tolerate one another's presence in the same feeding or breeding grounds. In some species, as among the large carnivores, they are solitary prowlers. This inability of males in nature to cooperate with one another is a serious handicap so far as the development of group ties is concerned.

The females, on the other hand, thanks to their maternal functions, are not handicapped in this manner. They form into broods composed of the mother and offspring in which cooperation exists and filial ties have a chance to develop. In some species, such as the primates, or even in a pride of lionesses, a number of females and offspring band together in a larger brood. Moreover, while the male animal has only himself to consider in the struggle for survival, the female through her maternal functions must provide for and protect her offspring as well as herself. Through the constant exercise of these group functions, it is normally the female, not the male, that is more intelligent, sagacious, cunning, and capable. This is recognized by hunters who regard the female, particularly those with cubs, as the more dangerous sex and take appropriate precautions. This keener wit in females is developed to its highest degree among the highest mammals—for it is here that the maternal functions and the care of offspring are most protracted—and reaches its apex with the primates. Even Robert Ardrey, an ardent partisan of male superiority, admits: "As the kingdom of the animal ascends, so likewise ascends the power of the female. . . . Masculine woollymindedness has been a source of female power for a long way back."[8] Robert Briffault more bluntly calls male animals more stupid than females.

These considerations show that there is no basis in nature for the "uterus theory" of female inferiority. If anything, nature favored the female sex since that is the sex upon which the perpetuation of the species pivots. Carrying out their maternal functions gave the females an advantage in the struggle for survival, enabling our branch of the anthropoids to pass from nature's mode of survival to the human mode of survival through labor activities. In the transition from ape to human it was the female, not the male, that led the way. Already more developed in her capabilities and capacity for cooperation, it was females who began productive life and thereby founded the new and unique human species.[9]

That is why, out of the maternal brood in the animal world, there arose the maternal clan system or "matriarchy" in the ancient human world. It is only in patriarchal class society, which came a million years after the birth of the human species, that woman was reduced to an animal-like level, forced to preoccupy herself with her maternal functions at the expense of the higher human values developed in the course of social life. In a society founded upon private property, the family institution, and male supremacy, woman's natural endowment—her uterus and maternal functions—were turned into the chains of exploitation and oppression she endures today. But this is a situation made by man, not by nature.

Those who subscribe to the "uterus theory" of female inferiority often try to prop up their false conclusions about women with an equally false theory about the everlasting superiority of the male sex. Reducing the science of biology to science fiction, they project the image of the patriarchal family of our times back into the animal world. To them the animal "family," like the human family, has male at its head, providing for and protecting his dependent wife and children, and this is what makes him superior. This animal hero is usually called the "dominant male." As portrayed by the fiction writers, he is the animal counterpart of the husband and father in patriarchal society. The more fanciful even portray this male animal as a kind of princely potentate, surrounded by a harem of wives, concubines, and female slaves, controlling their lives and destiny. What is the truth behind this fantasy?

The 'dominant male': fact and fiction

The phenomenon called the "dominant male" does exist in nature since animal males, as previously stated, are highly competitive and combative against one another. In the sexual realm each strives to gain first place by eliminating rivals. The animal that wins becomes dominant over the other males, at least for a time or until he himself may be displaced by a stronger male.

But the main point about this fight for dominance is usually left out or distorted. It is a struggle among males, each fighting the others. Even after the dominant animal has eliminated his rivals, this does not make him dominant over the female or group of females to which he gains access. So far as the females are concerned they may accept the winning male as their stud but that is all. Even this acceptance ends when the females enter their maternal cycle, at which time they retire from the orbit of all males to preoccupy themselves with giving birth and caring for their offspring. Whatever the outcome of the struggle among males, the females remain entirely self-sufficient and provide for their offspring without assistance from males.

Contrary to all the children's stories on the subject, often written by men who call themselves scientists, there is no such thing as a father-family in the animal world. Among some bird and fish species, males may participate in the care of the eggs. This does not make them families but rather a specialized form of procreation. In the great majority of species, above all the mammals which are in the direct line of human ascent, it is the mothers alone who perform all the functions connected with the care of offspring. As Briffault emphasizes, "Every adult animal, male or female, fends for itself as far as regards its economic needs" and the only exception is the provision made by the mother for her young.

In other words, male sexuality in the animal world does not bring about fatherhood functions; on the contrary, the combativeness connected with male sexuality is a hindrance in the development of such functions. It is only in the human world that we find a fully developed male counterpart to maternity which we call paternity. This came about when males began to emancipate themselves from direct biological controls—or instincts—and cultivated new and human traits. It was in and through social life that they learned a new kind of sexual behavior and subsequently acquired paternal functions.

It is sometimes said or implied that because males are the combative sex they are the "protectors" of their animal families. This too is a fiction. Among some primate species a periphery of males circles around a central core of females and offspring and in an indirect way furnishes an outer group of "sentinels" who sound an alarm in the event of danger. But male animals do not fight to protect their wives and offspring. They fight in defense of their own lives.

In the animal world every animal defends itself, either by fight or by flight. The sole exception to this rule is the female animal, who will fight to defend her offspring. Thus the so-called animal family is no more than a female brood, provided for and protected by the mother. It does not have the slightest resemblance to the patriarchal family in our society where the father provides for and rules over his wife and children.

Another familiar argument supposed to prove the natural superiority and dominance of the male sex over the female sex is based on the fact that in some species (although by no means all) the males are larger than the females or have more developed muscle power. There can be little doubt that the combative traits of males contributed to this extra musculature. As Henry W. Nissen of the Yerkes Laboratories said about primate males: "The bigger animal gets most of the food, the stronger male most of the females."[10]

But it is wrong to assume that this extra musculature represents a superiority of males over females; it is only a superiority of stronger males over weaker males. In nature it is the females who determine whether or not they wish to admit a male into their midst, and this holds true of the stronger males as well. When such admission takes place it is only during his good behavior and so long as the females find his presence convenient. This is borne out by the fact that when a female retires, as she does when she gives birth, she is left entirely alone by the males.

It is therefore a gross misrepresentation of animal life and behavior to portray the female as a helpless, dependent creature that cannot survive without the provision and protection of a "dominant male" playing the part of husband and father. The father-family is exclusively a human institution which, moreover, came into existence very late in social history, coincident with the development of private property and class divisions. Thus the myth of the animal "father-family" goes hand in hand with the "uterus theory" of female inferiority. The true facts about biology are distorted and falsified for the sake of concealing the social roots of female oppression.

Let us turn now to examine the way that anthropology is distorted to buttress a falsified biology. I call this—

The 'hunting theory' of female inferiority

This theory bases itself upon the first division of labor between the sexes which is usually described as follows: men were the hunters and warriors while women were the food gatherers and did the chores around the camp or home. Man's work of hunting is, of course, portrayed as by far the most important work, while woman's labor is considered inferior. Due to the handicaps they suffered from being born with a uterus, they had to stay at the campsite or dwelling place to nurse the children.

This stands the real situation on its head. The most important work in the primitive division of labor was not done by the male hunters but by the so-called stay-at-home women. Let us start with the food supply, the first and most basic requirement since people must eat before they can do anything else. It was the women gatherers and not the male hunters who provided the most stable and ample supplies. During the period when hunting was still precarious and men often returned to camp empty handed, the community's hunger was satisfied by the food collected by the women. In addition it was the women who gained control over the food supplies, not only by preparing them for today but preserving stocks for tomorrow. Women were the mainstay of the primitive commune.

But this was only the beginning of woman's work. There is no need to dwell here upon the enormous labor record of primitive women.[11] While men were occupied in seeking out animals as game food, it was the women who carried out the diverse forms of production from leather-making, pot-making, and handicrafts of all kinds to construction, medicine, and the development of the earliest forms of science. While one branch of woman's work, soil cultivation with the digging-stick, led to agriculture, another branch, the taming of wild animals, led to the raising of stock animals. These major advances not only laid the foundations for civilization but liberated men from hunting to participate in—and eventually take over—these higher forms of production. Thus it was not the men hunters but the women producers, proto-scientists, nurses, teachers, and transmitters of the social, cultural, and technical heritage who did the most important work in the first sexual division of labor.

The great mistake made by those who are blinded by the assumed superiority of the male sex is to overlook this broad social production of primitive women and view them as mere homebodies, serving a little family circle. There were no isolated, shut-in, private family households in the primitive clan system, just as there was no propertied ruling class to reduce women to family servitude as their share of labor. The primitive "households" were the pivot of a communal life and represented the earliest factories, laboratories, medical centers, schools, and social centers. The women of the matriarchal commune, working collectively, had not the slightest resemblance to their descendants today, each one puttering around in a little stump of a household.

This is not to denigrate the skills and techniques that men developed in their occupation of hunting. It is simply to redress the balance and put men's work in its proper place and perspective. Indeed, not only has woman's work been slighted; even man's work of hunting has not received a fully rounded appreciation. The most important aspect of the human hunting band was not connected with man's capacity for brute force or even with increasing the food supply as such. It was the qualitative advance made by men over the animals in achieving a working collaboration with one another.

It is often pointed out that hunting requires both strength and skill if man is to master the large and dangerous animals, and this is true. What is seldom mentioned, however, is the more important aspect, that men had to overcome their former animal nature, their rivalry, separatism, and individualism, to be able to band together in the human hunting group. They had to transform their competitive, combative animal relations into close-knit, cooperative human relations. The superiority of the human hunting band over any animal pack comes from the unbreakable principle that men hunting together must never under any circumstances hunt or kill one another. This exclusively human relation does not exist in the animal world. Thus, even in the matter of increasing the food supply of the community, it was only when men learned how to form the cooperative hunting band that this aim could be achieved.

How was such an impressive change brought about? All the evidence points to the collectivist society created by the clan mothers, which assimilated the men as clan brothers. As Robert Briffault writes on this point:

> In human societies there always exist means of establishing understandings and guarantees, and there are bonds of fellowship and brotherhood which are absent and impossible among animals.

> Hence primitive humanity, owing to its social character, is not under the same necessity to secure the satisfaction of its sexual instincts by sheer competitive struggle. . . . Animals tear their closest associates and even their sexual mates to pieces in the struggle for food; the member of the rudest and most primitive social group will starve rather than not share his food with his fellow-members. . . . So likewise in no human society, however primitive, is a lawless scramble for the possession of females to be found.[12]

It required a communistic society, which provided for the needs of all its members on an equal basis, to bring men together as cooperators who had formerly been separated and hostile to one another as animals. In that society men as well as women did their share of the work according to the division of labor that the primitive peoples themselves found to be most practical under the given conditions of life at that stage of their socioeconomic development. Many writers have downgraded the real worth of the work done by women, while glorifying man's work of hunting. The archaeologist, Grahame Clark, for example, sees women as lowly beings because "like their simian forebears" they are merely food-gatherers, while he refers to the "resplendent figure of Man the Hunter, prototype of Man the Warrior," as the great and superior sex.[13] This is male bias.

Elman R. Service, the Michigan anthropologist, takes a similar although more restrained view of the matter. He thinks males were the hunters not only because they "were probably stronger, swifter, and more combative, but, more importantly because females are so frequently handicapped by pregnancy and care of offspring."[14]

We can accept the deduction that the combative characteristics of males made them adaptable to hunting. But we must reject the conclusion that females were incapable of hunting because they were biologically handicapped by their uteruses. One has only to observe the behavior of the carnivores, the hunting animals, to see the fallacy of this argument, since the females are just as swift and skillful hunters as the males. There is no uterus handicap imposing inferiority upon lionesses and tigresses.

To be sure, the human species arose out of the food-gathering primates, not the hunting carnivores. But women who want to challenge the "hunting theory" of female inferiority are not obliged to unravel the whole complex of reasons why women were not the hunters in the first division of labor. It is sufficient to show the vastly superior amount of work and types of work done by the women as compared to the main occupation of men, the hunters. The exclusion of *one* occupation—for whatever reasons—only signifies that the women left this one out of their multiplicity of labor activities.

In the end, then, the "hunting theory" of female inferiority is just as absurd and untenable as the "uterus theory" from which it is derived. The one is a distortion of anthropology as the other is of biology. Yet these furnish the pseudoscientific platform under the propaganda that women have always been the inferior or second sex.

Have women always been oppressed?

Since the rise of the women's liberation movement some women writers and even anthropologists have become so influenced by these unscientific propositions that they have drawn a very pessimistic conclusion. Women, they say, have been the oppressed sex not simply under patriarchal society but throughout all human history. According to this view, if women were not subjugated to their husbands and fathers as they are in patriarchal nations, then they were under the thumb of their brothers or uncles in primitive communities. This can be called the "avunculate theory" of female oppression. What is the truth of the matter?

There are a number of primitive communities scattered around the world where old matriarchal practices and customs survive to a greater or lesser extent. These are usually called "matrilineal" communities because the line of kinship and descent is still traced through the mothers alone. But the matter goes deeper than this. In such regions the father-family is still poorly developed. A man may be recognized as the husband of the mother and yet not be recognized as the father of her children, or, if recognized, has only an extremely tenuous connection with them. As this is usually expressed, the children belong to the mother and her kin.

This means that the children belong not only to the mothers but also to the brothers of such a matrilineal community. In other words, the mothers' brothers, or maternal uncles, still perform the functions of fatherhood for their clan sisters' children that in patriarchal societies have been taken over by the father for his wife's children. For this reason such a community is sometimes called "the avunculate." The term "avunculate" refers to the mother's brother as the term "patriarch" refers to the father.

These matrilineal communities are survivals from the matriarchal epoch and, however much they have been altered since the patriarchal takeover, testify to the priority of the earlier social system. In fact, by the time anthropology began in the last century, most primitive clans had already become altered in their composition to a certain degree. Pairing couples, or what Morgan called "pairing families," had made their appearance in communities that had formerly been composed solely of clan mothers and brothers (or sisters and brothers).

But the pairing family, which was still a part of the collectivist maternal clan system, was a totally different kind of family than the patriarchal family which came in with class society. A new man from outside the clan was added to the maternal group—the husband of the woman who became his wife. However, while the husbands participated in providing for their wives and children, so long as the clan system prevailed the husbands remained subordinate and even incidental to the mothers' brothers. The mothers' brothers remained the basic economic partners of their clan sisters and guardians of their sisters' children.

Field anthropologists who reject the historical approach are caught in a serious dilemma when they encounter such primitive clan communities. For instance, Malinowski, in his studies of the Trobriand Islanders, describes these people and their "Principles of Mother-Right" as follows: "We find in the Trobriands a matrilineal society, in which descent, kinship, and every social relationship are legally reckoned through the mother only, and in which women have a considerable share in tribal life, even to the taking of a leading part in economic, ceremonial, and magical activities . . ."[15]

But because "these natives have a well-established institution of marriage," that is, cohabit as pairing couples, Malinowski goes through a tortuous search for the father.

But the mother's husband has not yet developed into a father in the true sense of the term. According to the natives themselves, the *tama,* which Malinowski insists upon calling the "father," is no more than "the husband of my mother." In some instances he is not even that; he is a *tomakava,* a "stranger," or, as Malinowski puts it, "more correctly an 'outsider'."[16] In other words, the man from "outside" the clan, who has achieved recognition as the husband of the mother in some places, still falls short of achieving his true father status.

There is a man, however, who performs the functions of fatherhood for his sister's children, in particular for her male children. That is the mother's brother. Malinowski writes:

> Social position is handed on in the motherline from a man to his sister's children, and this exclusively matrilineal conception of kinship is of paramount importance . . . people joined by the tie of maternal kinship form a closely knit group, bound by an identity of feelings, of interests, and of flesh. And from this group, even those united to it by marriage and by the father-to-child relation are sharply excluded. . . .[17]

Malinowski significantly observes what he calls the "twofold influence" or "duality" that permeates this matrilineal community as a result of matrimony impinging upon matriliny. Male children look to, and feel divided between, two adult men connected with the mother. On the one hand is the old-established mother's brother; on the other hand there is the newcomer, the mother's husband. What Malinowski doesn't bring out is that the Trobriand Islanders represent a matrilineal community in transition to patrilineal forms.

The pioneer anthropologists of the last century found many examples of matrilineal communities passing over to patrilineal and patriarchal forms of social organization. As E. Sidney Hartland sums up the evidence, patriarchal rule "made perpetual inroads upon mother-right all over the world; consequently matrilineal institutions are found in almost all stages of transition."[18]

The position of women in some of these surviving communities-in-transition remained largely unaltered, and they continued to enjoy economic independence and social esteem. In other regions, however, particularly those in which class relations, patriarchalism, and male supremacy have been superimposed upon a rude economy, women became as degraded as their sisters in class society. In such regions women can be as much oppressed by their brothers as by their husbands and fathers.

Australia is often offered as proof of the debased condition of primitive women. But, according to Spencer and Gillen, the highest authorities on the central tribes, there is a "great gap" between the old traditional period and the present. They conclude that the women formerly occupied a far different and higher position than in recent times.[19]

Robert Briffault, summing up this and other reports, maintains that patriarchalism, male domination, and the debased condition of women are "features of comparatively late origin" and have supplanted a former condition of female influence and esteem. "The Australian natives are not only a primitive, they are in many respects also a degraded race," he says, and that is why male domination, once instituted, proceeded to "its extreme consequences."[20] This should not be surprising in a continent where, through disease and other causes, the aboriginal population of 500,000 was reduced to 50,000 within a century after the coming of the white man.

In sharp contrast, there are many regions in which matriarchal customs have been preserved and there is no such debasement either of the women or of the men. Such examples can be found among the North American Indians, where male supremacy and oppression of women were nonexistent until they were brought over, along with whiskey and guns, by the civilized settlers from Europe. Briffault cites the following from the missionary J.F. Lafitau:

> Nothing is more real than this superiority of the women. It is in the women that properly consists the nation, the nobility of blood, the genealogical tree, the order of generations, the preservation of families. It is in them that all real authority resides; the country, the fields, and all the crops belong to them. They are the soul of the councils, the arbiters of war and peace.[21]

According to Alexander Goldenweiser, woman's influence was paramount in the election of chiefs. The activities of these chiefs were carefully watched and supervised by "the matrons," especially in questions of war, and if found unsatisfactory in any respect the dissatisfaction of the women brought about the deposition of the chiefs. As late as in the period of the Iroquois confederacy, he says, "women were more influential than men both in the election of chiefs and in their deposition . . . public opinion was more significantly that of the women than of the men in the group." Many a devastating war, he adds, "must have been averted by the wise counsel of the matrons."[22] The reality of woman's power is evidenced by the fact that the deeds of land transfer of the colonial government nearly all bear the signatures of women.

One of the most interesting confrontations between the Iroquois men and the white men who looked down upon women as the inferior sex is cited by Briffault. The chosen orator of the Iroquois, "Good Peter," addressed Governor Clinton in this manner to explain the high esteem of the Native Americans for women:

> Brothers! Our ancestors considered it a great offence to reject the counsels of their women, particularly of the Female Governesses. They were esteemed the mistresses of the soil. Who, said our forefathers, brings us into being? Who cultivates our land, kindles our fires, and boils our pots, but the women? Our women, Brother, say that they are apprehensive. . . . They entreat that the veneration of our ancestors in favour of the women be not disregarded, and that they may not be despised: the Great Spirit made them. The Female Governesses beg leave to speak with the freedom allowed to women and agreeable to the spirit of the ancestors. . . . For they are the life of the nation.[23]

Briffault also cites W.W. Rockhill, who remarked, "By what means have they made their mastery so complete and so acceptable to a race of lawless barbarians who but unwillingly submit to the authority of their chiefs, is a problem well worth consideration."[24]

These are hardly pictures of "eternally oppressed" women. The fact that some women in primitive regions became as oppressed as the women of patriarchal civilized nations does not prove that women have always been oppressed. All it proves is that in some regions, but not all, the degradation of the mothers and sisters also brought about the degradation of the mothers' brothers. Some mothers' brothers became as much male supremacists and oppressors of women as the patriarchal men who served as their models.

But historically, before the patriarchal takeover, there was no such thing as male supremacy over women—or contrariwise, of female domination over men. The clan community was communistic; a sisterhood of women and a brotherhood of men. The keystone of that social structure was equality in all spheres of life—economic, social, and sexual. Thus women were *not* always oppressed. The oppression of women began as an integral part of an oppressive society that overturned and supplanted the matriarchal commune. The "avunculate theory" of eternal female oppression is only a more sophisticated variation of the "uterus theory" of female inferiority. The one, like the other, must be rejected by women of the liberation movement.

Unfortunately, this has not been done by some influential writers such as Kate Millett. Although she is scornful of the proposition that biology is woman's destiny, this fighter on the side of women's liberation has been influenced by the antihistorical anthropologists. In her book *Sexual Politics* she writes that "both the primitive and civilized worlds are male worlds," and that women have always been oppressed, if not by patriarchal men then by men of the "avunculate."[25] Oddly enough she takes this position while admitting she does not know whether or not there was a matriarchal period.

Shulamith Firestone in her book *The Dialectic of Sex* plunges even deeper into the error of the eternal oppression of women. She recites the whole man-made litany on this score. According to Firestone, female oppression is older than recorded history; it goes all the way back "to the animal kingdom itself." Because of female biology, she says, productive work was beyond women's strength—showing her ignorance of the extensive labor record of primitive women. Again, says the author, because of her biology woman "remained in bondage to life's mysterious processes." Firestone is merely repeating the male designations for breeding and baby tending. Thus, she concludes, women have been "at the continual mercy of their biology" which has made them "dependent on males," whether these were clan brothers or husbands and fathers.

Firestone has fallen hook, line, and sinker for the "uterus theory" of female inferiority. Sweeping aside Marx and Engels who, she says, knew "next to nothing" about women as an "oppressed class," she maintains that "it was woman's reproductive biology that accounted for her original and continued oppression, and not some sudden patriarchal revolution."[26] Firestone, the feminist, parrots the antifeminist theme that "biology is woman's destiny" without bothering to critically examine the facts.

It is unfortunate that even some women anthropologists have made similar errors despite their studies of the subject.

Influenced, or perhaps intimidated, by the male-supremacist and bourgeois ideology which permeates anthropological circles, they too subscribe to the myth of the everlasting inferiority and oppression of women. The anthropologist Lucy Mair states, "In the simplest societies, and indeed in some industrialized ones, women are never wholly independent. . . ." They have always had to depend on males, whether brothers or husbands and fathers.[27] This sweeping statement is not even true of some matrilineal survivals in recent times where women retained their economic independence and social esteem. It was not true at all for the matriarchal epoch of social organization before male supremacy was born.

Kathleen Gough Aberle, of Vancouver, made the best contributions to the book *Matrilineal Kinship,* published in 1961 on the one-hundredth anniversary of Bachofen's *Das Mutterrecht* (Mother-Right). Yet she, too, thinks that women have always been oppressed. In a recent article written for the women's liberation movement she states, "The power of men to exploit women systematically springs from the existence of accumulated wealth" backed by the power of the state. This adheres to the Marxist viewpoint. But then she departs from the method of historical materialism when she says, "Even in hunting societies it seems that women are always in some sense the 'second sex,' with greater or lesser subordination to men."[28]

While this may be true of some hunting communities that have become altered in recent times, it was not true of the original hunting communities that existed in the period of the matriarchal commune. Let me emphasize: It was not the occupation of hunting that gave men superiority over women—it was the introduction of private property, class divisions, and the patriarchal family that brought about male supremacy and the oppression of women.

Social versus family division of labor

This brings us to the final point in the tangle of myths aiming to prove that women have always been the second sex. This one concerns the distinction between the primitive and civilized division of labor between the sexes. According to the prevailing propaganda, the division of labor between the sexes has always been the same, with woman's work confined to home and family. From the very beginning of human history to the present day the division of labor between the sexes is believed to have been a division between the husband and wife of a family. The husband goes out to work while the wife stays at home to take care of the household and children. Some women in the liberation movement are indignant because the husband gets paid for his work while the wife does not. But the injustice goes deeper than this. It involves the stunted, dependent, culturally sterile life of a woman caged up in a domestic enclosure doing stupefying chores.

Women are deprived of the kind of socialized work which would give them economic independence; such work is largely reserved for men. Marriage and the family are upheld as the finest career a true woman can pursue. Reactionary contraception and abortion laws force women to bear children whether they want to or not, and in the absence of child-care centers each individual woman is saddled with the burden of raising the children herself.

According to the churches and the guardians of the established order, women's place is in the home serving a husband and children, because the family has always existed. But it is not true that procreation, which is a natural function, is identical with the family, which is a man-made institution. While women have always been the procreators of children, they have not always been isolated in self-enclosed units, each woman serving a husband and family. The "eternal family" hoax is only the ultimate expression of the "uterus theory" of female inferiority.

The first division of labor between the sexes was not, as it is today, a division between a husband and wife, with the man doing outside work while the wife stayed at home doing housekeeping chores. Both sexes in primitive society performed social labor. This was possible because their system of communal production was accompanied by communal child care and education. Female children were trained by the adult women into their future occupations while the male children at a certain age were turned over to the adult men who became their tutors and guardians. Both production and child raising were originally social functions, performed by both women and men. It was only with the downfall of the matriarchal commune and its equalitarian relations between the sexes that women were dispossessed from social production and put into family servitude. Men took over in the new divisions of labor.

Historians often point out that with the advent of the new economy founded upon agriculture and stock raising, many new divisions of labor came into existence, replacing the former sexual division of labor. To give a few examples, pastoral activities became separated from farming; metallurgy, house construction, shipbuilding, textiles, pottery, and other crafts became specialized trades. Along with these divisions of labor in the crafts, there grew up specializations in the cultural sphere, from priests and bards to scientists and artists.

The roles of the sexes were radically transformed in the process. As these new divisions and subdivisions of labor grew and proliferated they became more and more—and finally exclusively—in the hands of the men. The women were squeezed out of these fields of social and cultural work—and pushed into home and family life. With the rise of state and church power, women were taught that their whole lives were bounded by the four walls of a home and the best women were those who served their husbands and families without complaint. In this elevation of men and downgrading of women, they were compelled to forfeit not only their former place in social production but also their former system of communal child care.

To be sure, women of the plebeian classes, the "common people," have always worked. In the long agricultural period they worked on farms as well as in cottage crafts, and they did all this along with bearing children and taking care of households. But working in and through and for an individual husband, home, and family, is by no means the same thing as engaging in socialized labor in a communal society. Participation in social production develops the mind and body; isolation and preoccupation with home chores weakens them and narrows the outlook.

In other words, the division of labor between the sexes has *not* always been the same. The male-dominated division of labor that came in with class society, private property, and the patriarchal family represented a colossal robbery of the women. This is even more true today with the reduction of the extended, productive farm family to the tiny, nuclear, consuming family of the urban era.

To refute the myths that have helped to keep women oppressed—from the "uterus theory" to the "eternal family" propaganda—is not simply a matter for scientific and historical correction. It has profound implications for the women's liberation movement. The argument that woman's biological makeup is responsible for her social inferiority is the chief stock-in-trade for the male supremacists. If this claim proves to be unfounded their position collapses.

Females in nature suffer no disabilities compared to males as a result of their biology. Nor were women downgraded as a result of their maternal role in preclass society. They were held in the highest esteem for their combined functions as producer-procreatrix. Woman's position in society, therefore, has been shaped and reshaped by changing historical conditions. The drastic transformation that overturned matriarchal communism brought about the downfall of the female sex. It was with the rise of patriarchal class society that the biological makeup of women became the ideological pretext for justifying and continuing the dispossession of women from social and cultural life and keeping them in a servile status.

Only by recognizing this can women come to grips with the real causes of our subjugation and degradation which are today bound up with the structure of the capitalist system. Our struggle for liberation will be hindered so long as we are hoodwinked into believing that nature rather than this society is the source of our oppression.

A banner carried by women in a recent demonstration proclaimed, "Biology Is Not Woman's Destiny." This should become a watchword of the feminist movement.

NOTES

1. Frederick Engels, *The Origin of the Family, Private Property, and the State* (New York: Pathfinder Press, 1972). Also included is Engels's essay "The Part Played by Labour in the Transition from Ape to Man."

2. *Scientific American,* September 1960.

3. V. Gordon Childe, *What Happened in History* (Harmondsworth, Middlesex: Penguin, 1960), 27.

4. E. Adamson Hoebel, *Man in the Primitive World: An Introduction to Anthropology* (London and New York: McGraw-Hill, 1949), 92.

5. Karl Marx, *A Contribution to the Critique of Political Economy* (Chicago: Charles H. Kerr, 1904), 279.

6. *Scientific American,* September 1960, p. 77.

7. Solly Zukerman, *The Social Life of Monkeys and Apes* (London: Routledge and Kegan Paul, 1932), 69.

8. Robert Ardrey, *African Genesis: A Personal Investigation into the Animal Origins and Nature of Man* (New York: Dell, 1963), 125.

9. See "The Myth of Women's Inferiority" in Evelyn Reed's *Problems of Women's Liberation* (New York: Pathfinder Press, 1970).

10. *Scientific American,* September 1960.

11. See "The Myth of Women's Inferiority."

12. Robert Briffault, *The Mothers: A Study of the Origins of Sentiments and Institutions* (London: Allen and Unwin, 1952), vol. 2, p. 118.

13. John Grahame Douglas Clark, *From Savagery to Civilization* (London: Corbett, 1946), vol. I, p. 8.

14. Elman R. Service, *Primitive Social Organization* (New York: Random House), 39.

15. Bronislaw Malinowski, *The Sexual Life of Savages in North-Western Melanesia* (New York: Harcourt Brace, 1929), 3.

16. Ibid., 5, 6.

17. Ibid., 4.

18. E. Sidney Hartland, *Primitive Society: The Beginning of the Family and the Reckoning of Descent* (London: Menthuen, 1921), 34.

19. Baldwin Spencer and F.J. Gillen, *The Native Tribes of Central Australia* (London: Macmillan, 1889), 195–96.

20. Briffault, vol. 1, pp. 338–39.

21. Ibid., 316.

22. Alexander Goldenweiser, *Anthropology: An Introduction to Primitive Culture* (London: Crofts, 1937), 365.

23. Briffault, vol. 1, pp. 316–17.

24. Ibid., 327.

25. Kate Millet, *Sexual Politics* (New York: Doubleday, 1969), 46, 25.

26. Shulamith Firestone, *The Dialectic of Sex: The Case for Feminist Revolution* (New York: William Morrow, 1970), 2, 82, 83.

27. Lucy Mair, *An Introduction to Social Anthropology* (New York and Oxford: Oxford University Press, 1970), 83.

28. *Up From Under,* January–February 1971.

WOMEN'S LIBERATION AND SOCIALISM

Abortion Is a Woman's Right!
Pat Grogan, Evelyn Reed

Why abortion rights are central not only to the fight for the full emancipation of women, but to forging a united and fighting labor movement.
$6. Also in Spanish.

Cosmetics, Fashions, and the Exploitation of Women
Joseph Hansen, Evelyn Reed, Mary-Alice Waters

How big business plays on women's second-class status and economic insecurities to market cosmetics and rake in profits. And how the entry of millions of women into the workforce has irreversibly changed relations between women and men—for the better.
$15. Also in Spanish and Farsi.

Women's Liberation and the African Freedom Struggle
From Matriarchal Clan to Patriarchal Family
Thomas Sankara

"There is no true social revolution without the liberation of women," explains the leader of the 1983–87 revolution in the West African country of Burkina Faso.
$8. Also in Spanish, French, and Farsi.

Communist Continuity and the Fight for Women's Liberation
Documents of the Socialist Workers Party 1971–86

How did the oppression of women begin? Who benefits? What social forces have the power to end women's second-class status? 3 volumes, edited with preface by Mary-Alice Waters.
$30

WWW.PATHFINDERPRESS.COM

Workers and the US rulers'

Three books for today's spreading and deepening debate among working people looking for a way forward in face of capitalism's global economic and social calamity and wars.

The Clintons' Anti-Working-Class Record
Why Washington Fears Working People

Jack Barnes

Hillary Clinton calls workers who refused to vote for her "deplorables." Donald Trump uses demagogy to try to turn working people against each other. Barnes documents US capitalism's drive for profits over the last quarter century and the consequences for working people, who want to "drain the swamp" of capitalist politics as usual.

$10. Also in Spanish, French, and Farsi.

deepening political crisis

Are They Rich Because They're Smart?

Class, Privilege, and Learning under Capitalism

Jack Barnes

Takes apart self-serving rationalizations by layers of well-paid professionals that their intelligence and schooling equip them to "regulate" the lives of working people, who don't know our own best interests.

$10. Also in Spanish, French, and Farsi.

Is Socialist Revolution in the US Possible?

A Necessary Debate among Working People

Mary-Alice Waters

An unhesitating "Yes"—that's the answer by Waters. Possible—but not inevitable. That depends on us.

$10. Also in Spanish, French, and Farsi.

WWW.PATHFINDERPRESS.COM

from Pathfinder

Malcolm X, Black Liberation, and the Road to Workers Power
JACK BARNES

Drawing lessons from a century and a half of struggle, this book helps us understand why it is the revolutionary conquest of power by the working class that will make possible the final battle against class exploitation and racist oppression and open the way to a world based on human solidarity. A socialist world.
$20. Also in Spanish, French, Farsi, Arabic and Greek.

Capitalism's World Disorder
Working-Class Politics at the Millennium
JACK BARNES

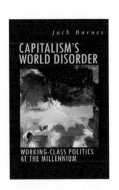

The social devastation and financial crises, the coarsening of politics, the cop brutality and acts of imperialist aggression accelerating around us—all are products not of something gone wrong with capitalism but of its lawful workings. Yet the future can be changed by the united struggle and selfless action of working people conscious of their power to transform the world.
$25. Also in Spanish and French.

The Working Class and the Transformation of Learning
The Fraud of Education Reform under Capitalism
JACK BARNES

"Until society is reorganized so that education is a human activity from the time we are very young until the time we die, there will be no education worthy of working, creating humanity."
$3. Also in Spanish, French, Farsi, and Greek.

WWW.PATHFINDERPRESS.COM

Cuba's Socialist Revolution

Women in Cuba: The Making of a Revolution within the Revolution
Vilma Espín, Asela de los Santos, Yolanda Ferrer

The integration of women in the ranks and leadership of the Cuban Revolution was inseparably intertwined with the proletarian course of the revolution from the start. This is the story of that revolution and how it transformed the women and men who made it.
$20. Also in Spanish and Greek.

The First and Second Declarations of Havana

Nowhere are the questions of revolutionary strategy that today confront men and women on the front lines of struggles in the Americas addressed with greater truthfulness and clarity than in these uncompromising indictments of imperialist plunder and "the exploitation of man by man." Adopted by million-strong assemblies of the Cuban people in 1960 and 1962.
$10. Also in Spanish, French, Farsi, Arabic, and Greek.

Cuba and the Coming American Revolution
Jack Barnes

This is a book about the struggles of working people in the imperialist heartland, the youth attracted to them, and the example set by the Cuban people that revolution is not only necessary—it can be made. It is about the class struggle in the US, where the revolutionary capacities of workers and farmers are today as utterly discounted by the ruling powers as were those of the Cuban toilers. And just as wrongly.
$10. Also in Spanish, French, and Farsi.

Cuba and Angola: The War for Freedom
Harry Villegas ("Pombo")

The story of Cuba's unparalleled contribution to the fight to free Africa from the scourge of apartheid. And how, in the doing, Cuba's socialist revolution was strengthened.
$10. Also in Spanish.

WWW.PATHFINDERPRESS.COM

EXPAND YOUR REVOLUTIONARY LIBRARY

Malcolm X Talks to Young People

"The young generation of whites, Blacks, browns, whatever else there is— you're living at a time of revolution," Malcolm said in December 1964. "And I for one will join in with anyone, I don't care what color you are, as long as you want to change this miserable condition that exists on this earth." Four talks and an interview given to young people in the last months of Malcolm's life. $15. Also in Spanish, French, Farsi, and Greek.

Our History Is Still Being Written
The Story of Three Chinese Cuban Generals in the Cuban Revolution

Armando Choy, Gustavo Chui, and Moisés Sío Wong talk about the historic importance of Chinese immigration to Cuba, and the place of Cubans of Chinese descent in more than five decades of revolutionary action and internationalism. $17. Also in Spanish, Farsi, and Chinese.

The Jewish Question
A Marxist Interpretation
ABRAM LEON

Traces the historical rationalizations of anti-Semitism to the fact that Jews became a "people-class" of merchants and moneylenders in the centuries before industrial capitalism. And how, in times of crisis, capitalists mobilize renewed Jew-hatred to incite reactionary forces and disorient working people about the true source of their impoverishment. $25. Also in Greek.

"It's the Poor Who Face the Savagery of the US 'Justice' System"
The Cuban Five Talk about Their Lives within the US Working Class

From police to courts to prisons to parole: how the US "justice" system works as "an enormous machine for grinding people up." Five Cuban revolutionaries—framed up by the US government and incarcerated for 16 years—draw on their own experience to explain the human devastation wrought by capitalist "justice." And what makes socialist Cuba different. $15. Also in Spanish, Farsi, and Greek.

Teamster Politics
FARRELL DOBBS

Tells how Teamster Local 544 organized the unemployed and truck owner-operators into fighting union auxiliaries. Deployed a Union Defense Guard to respond to the fascist Silver Shirts. Combated FBI frame-ups. Campaigned for workers to break politically from the bosses and organize a labor party based on the unions. And mobilized labor opposition to US imperialism's entry into World War II. $19. Also in Spanish.

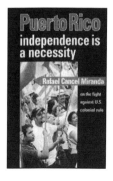

Puerto Rico: Independence Is a Necessity
RAFAEL CANCEL MIRANDA

One of the five Puerto Rican Nationalists imprisoned by Washington for more than 25 years speaks out on the brutal reality of US colonial domination, the campaign to free Puerto Rican political prisoners, the example of Cuba's socialist revolution, and the ongoing struggle for independence. $6. Also in Spanish and Farsi.

WWW.PATHFINDERPRESS.COM

PATHFINDER AROUND THE WORLD

Visit our website for a complete list of titles and to place orders

www.pathfinderpress.com

PATHFINDER DISTRIBUTORS

UNITED STATES
(and Caribbean, Latin America, and East Asia)
  Pathfinder Books, 306 W. 37th St., 13th Floor
  New York, NY 10018

CANADA
  Pathfinder Books, 7107 St. Denis, Suite 204
  Montreal, QC H2S 2S5

UNITED KINGDOM
(and Europe, Africa, Middle East, and South Asia)
  Pathfinder Books, 2nd Floor, 83 Kingsland High St.
  Dalston, London, E8 2PB

AUSTRALIA
(and Southeast Asia and the Pacific)
  Pathfinder Books, Suite 22, 10 Bridge St.
  Granville, NSW 2142

NEW ZEALAND
  Pathfinder, 188a Onehunga Mall Rd., Onehunga, Auckland 1061
  Postal address: P.O. Box 13857, Auckland 1643

Join the Pathfinder Readers Club
to get 15% discounts on all Pathfinder titles and bigger discounts on special offers.
Sign up at www.pathfinderpress.com or through the distributors above.
$10 a year